INVESTIGATING MYSTERIOUS PLACES

THE BERMUDA TRIANGLE

BLACK HOLE OF THE ATLANTIC

by Allan Morey

CAPSTONE PRESS
a capstone imprint

Published by Capstone Press, an imprint of Capstone
1710 Roe Crest Drive, North Mankato, Minnesota 56003
capstonepub.com

Library of Congress Cataloging-in-Publication Data is available on the Library of Congress website.

ISBN: 9781669093732 (hardcover)
ISBN: 9781669093688 (paperback)
ISBN: 9781669093695 (ebook PDF)

Summary: Travel to the Bermuda Triangle, where compasses spin wildly and time seems to stand still! Beyond the tales of vanished ships and airplanes, this book explores the unanswered questions that make the Bermuda Triangle a must-visit site for the curious and brave. From mysterious magnetic phenomena to underwater cities and even time warps, discover the science and the myths that surround the so-called Black Hole of the Atlantic.

Editorial Credits
Editor: Donald Lemke; Designer: Tracy Davies; Media Researcher: Svetlana Zhurkin; Production Specialist: Katy LaVigne

Image Credits
Dreamstime: Philcold, 19; Getty Images: Chris Clor, cover, back cover, 1, HadelProductions, 11, Heritage Images/Heritage Art, 24, Houston Chronicle/Brett Coomer, 27, John Lund, 6, Mark Stevenson, 21, Science Photo Library/Victor Habbick Visions, 7, 22; Shutterstock: Albert Stephen Julius, 25, Aliaksei Hintau (smoke background), 2 and throughout, Allexxandar, 16, Anton Balazh, 29, Everett Collection, 17, Gemini Pro Studio, 9, Harvepino, 13, Janice Carlson, 26, kikk, 18, Minerva Studio, 23, Peter Hermes Furian, 10, Premio Stock (triangle icon), cover and throughout, Sara Giordani Meurer, 5, zef art, cover (airplane); U.S. Naval History and Heritage Command: 15

TABLE OF CONTENTS

Chapter One

VANISHED!

What is the Bermuda Triangle? It is an area of the North Atlantic Ocean. It lies off the eastern coast of the United States.

But what makes this stretch of ocean so mysterious?

People tell many strange tales about the Bermuda Triangle. In some stories, ships vanish and are never seen again. In others, rescue crews find ships with no one aboard.

There are also stories about airplanes disappearing over the area. Many believe more than 50 ships and 20 planes have been lost in the Bermuda Triangle.

FACT

The Bermuda Triangle is also called the Devil's Triangle or Hoodoo Sea.

Chapter Two

NORTH ATLANTIC OCEAN

Where is the Bermuda Triangle?

It is not a **landmark** found on a map. It does not have any borders. The triangle does not even have an exact location people agree on.

The location of the supposed Bermuda Triangle

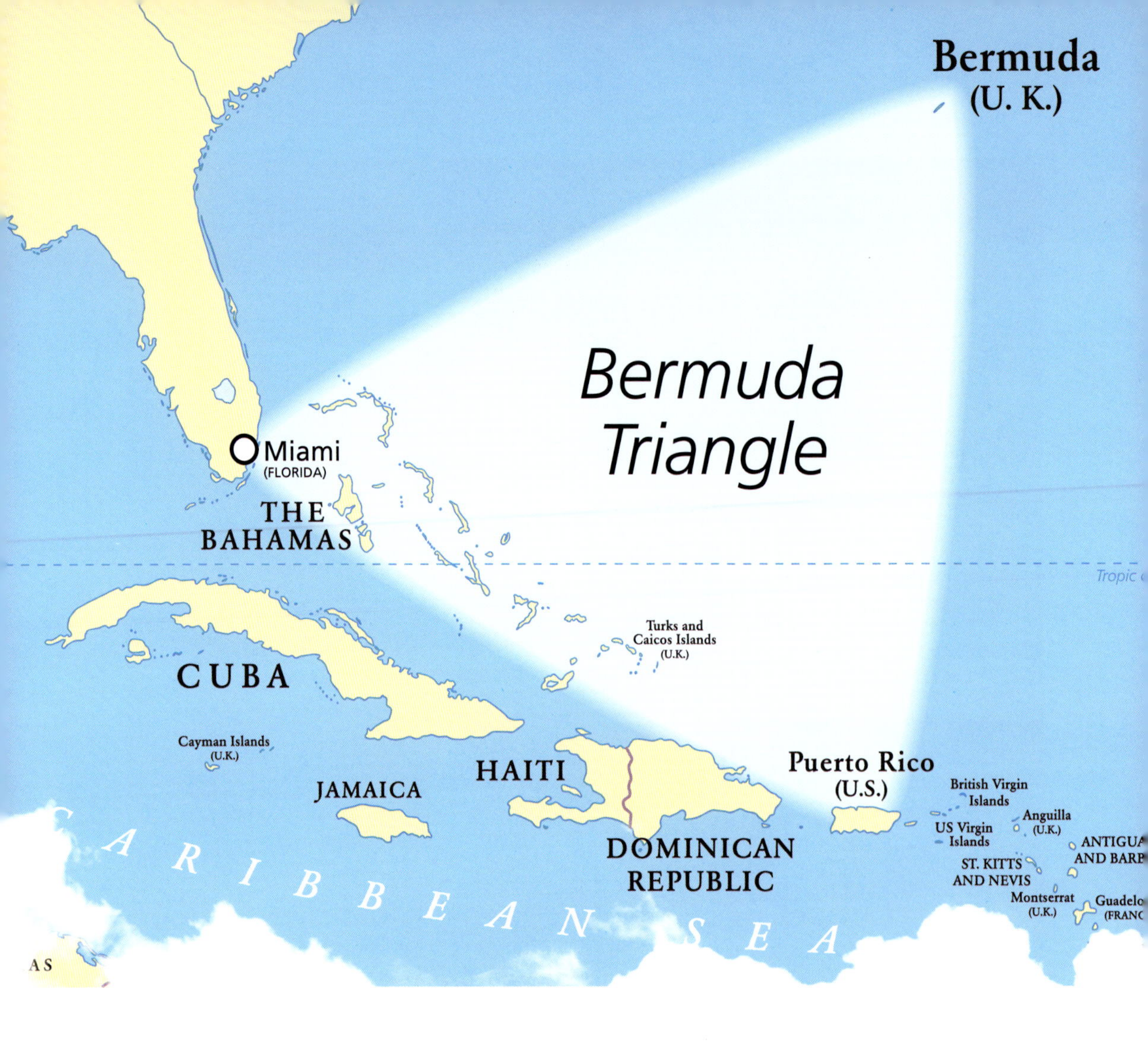

One corner is located off the southeast coast of Florida. Another is near the island of Puerto Rico. The third is close to the island of Bermuda. The area between these places roughly forms a triangle.

Weather in the Bermuda Triangle is **unpredictable**. Warm waters from the Gulf of Mexico flow through the area. As they move into cooler waters, storms often form.

Many **tropical storms** and **hurricanes** in the North Atlantic blow through the Bermuda Triangle. But does bad weather explain the many **disappearances**?

FACT

The Bermuda Triangle is between 500,000 and 1.5 million square miles (1.3–3.9 million square kilometers).

A hurricane swirls over the Bermuda Triangle.

Chapter Three

FAMOUS STORIES

One of the most famous stories about the Bermuda Triangle involves the U.S. Navy ship USS *Cyclops*. In 1918, it sailed north from Brazil to Baltimore, Maryland.

While cruising through the Bermuda Triangle, the ship mysteriously disappeared! The crew never sent a **distress** signal. All 306 crew members vanished with the ship.

The USS *Cyclops* supported battleships in World War I (1914–1918) before its disappearance.

So what happened to the USS *Cyclops*? Nobody knows. Wreckage from the ship has never been found.

Countless shipwrecks on the ocean floor hold many secrets.

U.S. military torpedo bombers flying in the early 1940s

In December 1945, five U.S. military planes were flying through the Bermuda Triangle. The **squadron** was on a training mission now known as Flight 19.

During the mission, one of the pilots radioed for help. He said that they were lost. A rescue plane went to search for the squadron.

The five planes were never found. The rescue plane also disappeared! Where did they all go?

That answer remains a mystery.

FACT

The airplanes that disappeared in 1945 were Avenger torpedo bombers used during World War II (1939–1945).

Chapter Four

WHAT'S THE TRUTH?

How do people explain the mysteries of the Bermuda Triangle? Everyone has many different ideas. Often, these ideas can be as strange as the mysteries themselves.

Many people believe the disappearances are **supernatural**. They think something out of this world is happening!

Maybe the disappearances were caused by aliens from outer space. Maybe giant **waterspouts**. Or even stranger, maybe the missing ships and planes traveled through time!

Some believe alien spaceships are to blame for the disappearances.

Tornado-like waterspouts can be dangerous to ships at sea.

Others believe the disappearances happened because of bad weather. To many, that is much more believable.

But there could be other reasons for the disappearances. The USS *Cyclops* vanished during World War I. The ship could have been sunk by an enemy submarine.

The airplanes of Flight 19 may have been lost due to human **error**.

A German submarine during World War I

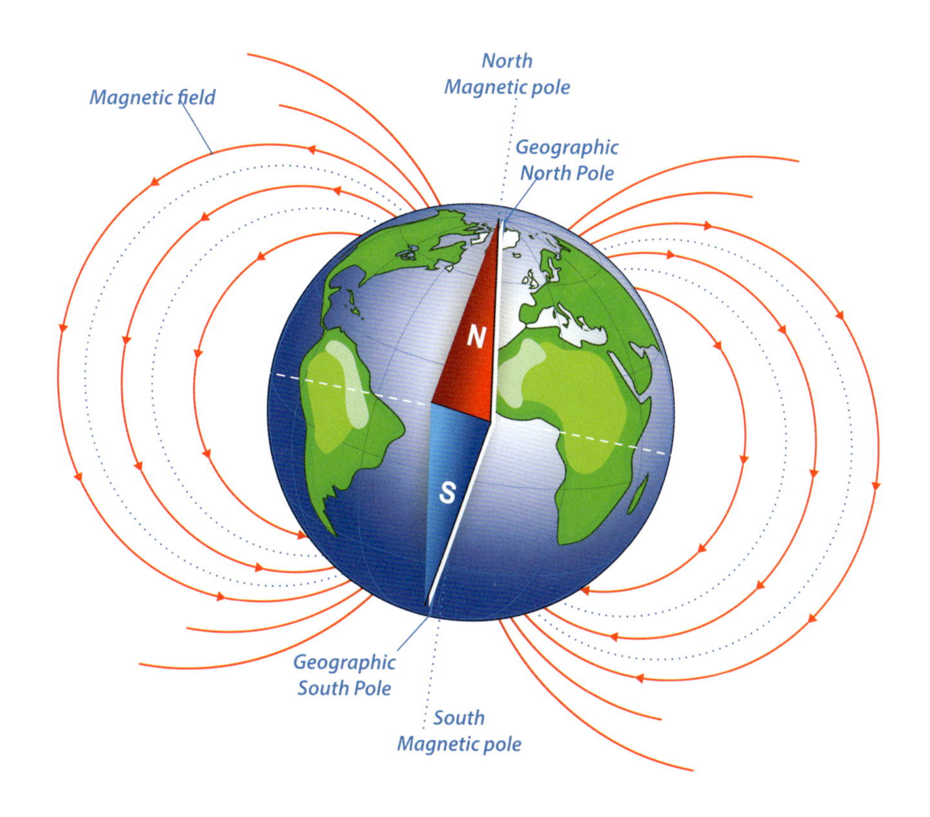

Earth is surrounded by a **magnetic field**. It makes compasses point north. Because of a weaker magnetic field in the Bermuda Triangle, compasses act differently. If a pilot misreads their compasses, they could get lost.

Then there are people who think there is no mystery at all. The Bermuda Triangle is a well-traveled part of the North Atlantic Ocean.

A cruise ship sails through the Caribbean Sea, an area not far from the Bermuda Triangle.

NOAA scientists monitor weather to keep ships safe at sea.

It is not surprising that some ships have sunk there. The National Oceanic and Atmospheric Administration (NOAA) claims there has not been an unusual amount of disappearances in the area.

But until we truly know what happened to the missing ships and airplanes, the Bermuda Triangle will remain a mysterious place.

FACT

Christopher Columbus reported seeing strange lights when he sailed through the Bermuda Triangle.

Glossary

disappearance (dis-uh-PEER-uhns)—the act of something or someone vanishing or going missing without explanation

distress (duh-STRES)—a state of danger or trouble

error (EHR-ur)—something produced by mistake

hurricane (HER-uh-kayn)—a swirling storm formed in the tropics with winds of 74 miles (119 km) per hour or greater

landmark (LAND-mark)—an object that marks the boundary of land

magnetic field (mag-NET-ik FEELD)—an invisible force field created by a magnet that exists all around it

squadron (SKWA-drun)—any of several units of military organization

supernatural (soo-per-NAT-er-uhl)—related to forces or phenomena beyond scientific understanding or explanation

tropical storm (TRAH-pih-kuhl STORM)—a storm with strong winds of over 39 miles (63 km) per hour

unpredictable (uhn-pri-DIK-tuh-buhl)—not able to be foreseen or anticipated

waterspout (WAW-ter-spout)—a funnel-shaped column of water that occurs over bodies of water, similar to tornadoes but composed of water vapor rather than air

Read More

Deniston, Natalie. *Bermuda Triangle.* Minneapolis: Jump!, Inc., 2025.

Harder, Megan. *Inside the Bermuda Triangle.* Minneapolis: Lerner Publications, 2023.

Troupe, Thomas Kingsley. *Searching for Bermuda Triangle Answers.* Mankato, MN: Black Rabbit Books, 2021.

Internet Sites

Britannica Kids: Bermuda Triangle
kids.britannica.com/kids/article/Bermuda-Triangle/598956

History.com: Bermuda Triangle
history.com/topics/folklore/bermuda-triangle

Science Kids: The Bermuda Triangle
sciencekids.co.nz/sciencefacts/earth/bermudatriangle.html

Index

About the Author

Some of Allan Morey's favorite childhood memories are from the time he spent on a farm in Wisconsin. Every day he saw cows, chickens, and sheep. He even had a pet pig named Pete. He developed a great appreciation of animals, big and small. Allan currently lives in St. Paul with his family and dogs, Stitch and Enzo, who keep him company while he writes.

LET'S DANCE!

LATIN DANCE

CHERRY LAKE PRESS
Ann Arbor, Michigan

by Joyce Markovics

Published in the United States of America by
Cherry Lake Publishing Group
Ann Arbor, Michigan
www.cherrylakepublishing.com

Reading Adviser: Beth Walker Gambro, MS Ed., Reading Consultant, Yorkville, IL

Content Adviser: Justin Wingenroth, dancer and owner/director of The Dance Conservatory, Cortland Manor, New York

Book Designer: Ed Morgan, Bowerbird Books

Photo Credits: © Andy-pix/Shutterstock, cover and title page; © JACK.Q/Shutterstock, 5; © Master1305/Shutterstock, 7; © Artur Didyk/Shutterstock, 8; © Lena Nester/Shutterstock, 9; © ooo.photography/Shutterstock, 10–11; © Roberto Galen/Shutterstock, 12; © Joma_JOMANOX/Shutterstock, 13; Wikimedia Commons, 14; © Oscar garces/Shutterstock, 15; © Celso Pupo/Shutterstock, 16; © catwalker/Shutterstock, 17 top; © Vladimir Gappovr/Shutterstock, 17 bottom; Dicklyon/Wikimedia Commons, 18; © Kobby Dagan/Shutterstock, 19; © MIRISCH-7 ARTS/UNITED ARTISTS/ Album/Newscom, 20; © JM11/Mandatory Credit: WENN/Newscom, 21; © Pacific Press/Sipa USA/Newscom, 22; © JM11/Joseph Marzullo/WENN/Newscom, 23; © Sthanlee Mirador/Sipa USA/Newscom, 25; © Kiselev Andrey Valerevich/Shutterstock, 27.

Cherry Lake Press is an imprint of Cherry Lake Publishing Group.

Library of Congress Cataloging-in-Publication Data has been filed and is available at catalog.loc.gov

Printed in the United States of America

Contents

It's SHOWTIME!

"Latin is the most passionate and emotional dance form."
—Susan Montero

Music with a marching beat fills the theater. A male dancer in a red and gold outfit lifts his arms. Holding his chest high, he stamps his feet. The dancer looks like a Spanish bullfighter called a matador. He peers at a female dancer across the dance floor. She has a long dress on. It flows like a matador's cape. She gathers her skirt and walks slowly toward the male dancer. Her chin is lifted as she locks eyes with him. They come together and spin around the room. The dancers separate and strike a pose. Then they continue dancing—and posing. The **intensity** of the dance increases. Suddenly, the male dancer drops to one knee, and his partner parades around him. Then the couple strikes one final dramatic pose. They are dancing the paso doble.

A couple dances the paso doble.

The term *paso doble* means "double step" or "two step" in Spanish. According to a Spanish legend, the movements of dance were inspired by a bullfight.

What Is LATIN DANCE?

"Do it big, do it right, and do it with style."
—Fred Astaire

The paso doble is one of dozens of Latin dances. Latin dances are known for their high energy, **passion**, and expressive music. Every dance has its own music and rhythms, which can be fast, slow, or a combination of both. The music is just as important as the dance steps! It guides the dancers' movements.

Besides the paso doble, other Latin dances include salsa, mambo, merengue, rumba, cha-cha, samba, and several others. Each dance has **signature** steps. The salsa, for example, is a party dance. It's known for fast footwork, turns, and a lot of movement in the hips. It has a "quick, quick, slow" rhythm. The merengue is related to the salsa but has its own flavor. To perform it, dancers use marching steps. They shift their weight from step to step.

Costumes for Latin dancers are designed to draw the eye. They're usually dramatic and sparkly!

Latin dances often involve partnering. This is when the dancers hold and face each other. However, in some styles, the dancers separate and then come back together.

The rumba, on the other hand, is a more **romantic** dance. It has slower hip movements. Rumba dancers sway their bodies from side to side. Then they make quick turns. Throughout the dance, dancers maintain eye contact.

Professional rumba dancers

People salsa dancing in a club

The mambo and cha-cha are fast Latin dances. They follow a quick three-step movement. The mambo has strong hip movements as well as speedy footwork. It looks like the salsa. But it is distinct. And the music is different too. The cha-cha—or cha, cha, cha—grew out of the mambo. It includes a series of small, fast steps. The dancers shift their weight from one foot to the other and appear to glide. The name of the dance comes from the sound the dancers' shoes make on the dance floor!

Latin dances can be traced to many different areas of the world. These include Africa, Europe, Central and South America, and the Caribbean.

Most people dance just for the fun of it. However, there are also **professional** Latin dancers. They perform on stage for audiences or in competitions. When these dancers compete, they show off their skills. Then they are judged on how well they perform. There are two main competitive schools of Latin ballroom dance. These are the International and American schools. Each school has its own dances and competitions.

Competitive ballroom dance is just one aspect of Latin dance. Folk dance is another. Folk dances reflect the lives of regular people from certain parts of the world. Many have roots that go way back. In fact, for thousands of years, people have been dancing. Let's dive into their remarkable stories and the history of Latin dance. We'll also explore how Latin dance became what it is today!

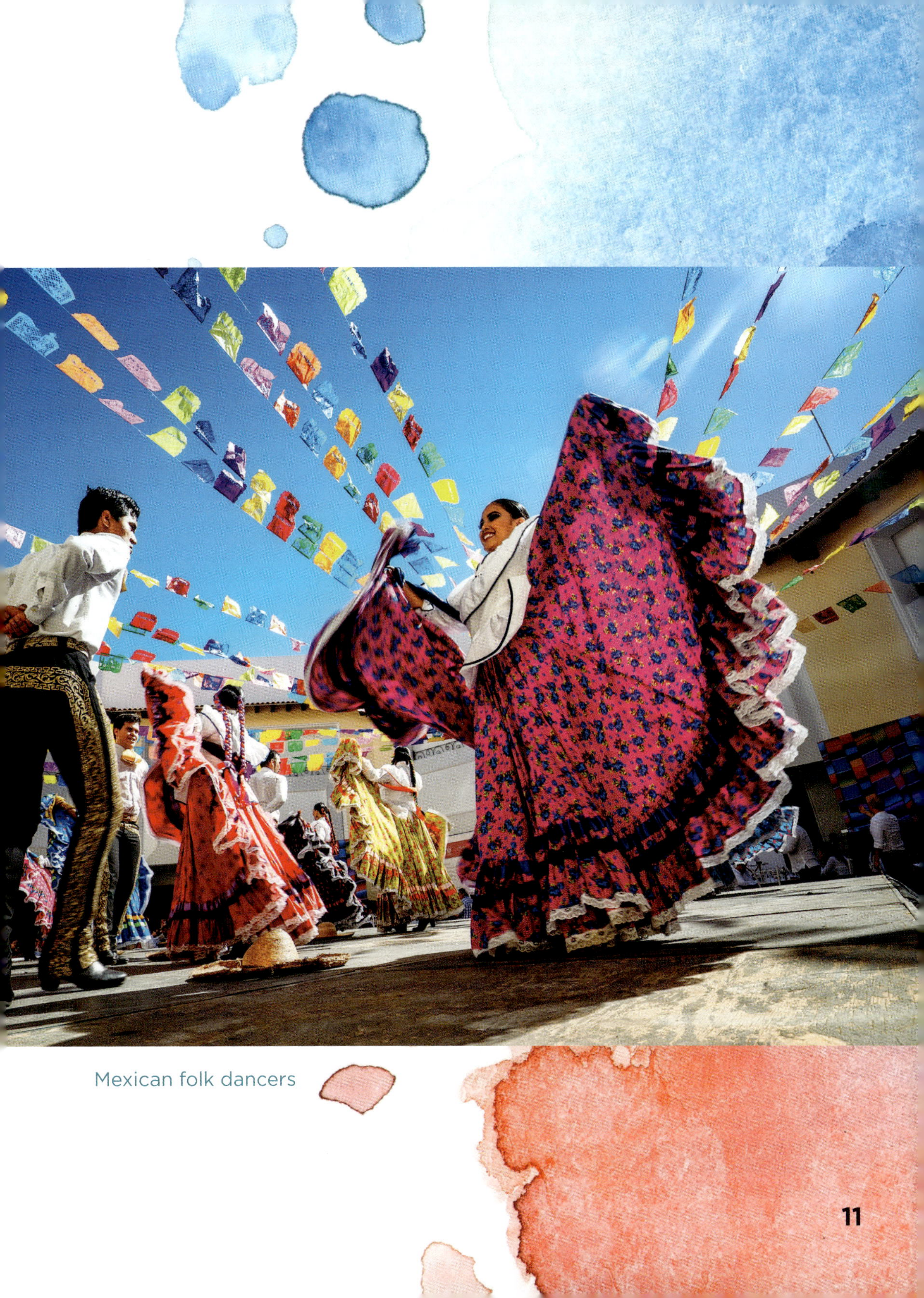

Mexican folk dancers

A Brief History

"Dance responds to community needs and realities."
—Susan de Leon, Aztec Dance Leader

Throughout Central and South America, early **Indigenous** peoples danced. For example, the Aztec people often danced in a circle to honor their gods. They wore colorful feather headdresses as they moved to the rhythm of drums. The dance represented a connection between body and spirit. Other Aztec dances showed battles or scenes of everyday life.

Today, people still dance in the tradition of the Aztecs.

Starting in the late 1400s, Europeans from Spain and Portugal arrived in the Americas. They wanted to find new trade routes to Asia. Instead, they encountered Indigenous peoples and their great **civilizations**. The European explorers wrote about the music and dance they saw. Yet they had no respect for the Indigenous peoples and their way of life. Soon, they **colonized** Central and South America. The European settlers killed and enslaved the Indigenous peoples and stole their land. They also enslaved African people, shipped them to the Americas, and forced them to work the stolen land.

The dance of the fliers may have been performed to end a drought.

Indigenous peoples from central Mexico, including the Totonac and Nahua people, developed the *danza de los voladores*, or "dance of the fliers." The dance involves four people swinging from the top of a very tall pole.

This art from the 1700s shows people dancing the fandango.

After colonization, the dances of the Americas changed. New ones developed. The enslaved Africans brought dances from their home countries with them. Many were performed to drum music and focused on hip and lower body movements. The Europeans had their own dances too. The fandango, for example, is an energetic partner dance from Spain and Portugal.

The cumbia is a dance that combines Indigenous, African, and European **cultures**. It began in Colombia in South America and later spread to other countries. Typically, it's a partner dance with side-to-side shuffling steps performed in a circle. It can also be danced as a group. Cumbia music has a signature sound. It's known for its heavy drums, airy flutes, and rattling maracas.

A cumbia musician shaking maracas

Maracas are instruments typically made from hollow gourds. They're filled with dried beans or stones and shaken to create sound.

Samba dancers wear elaborate and colorful costumes during Carnival.

Meanwhile, in Brazil, enslaved Africans created a dance called the samba to lively drum-based music. To perform it, dancers form a circle. One dancer at a time enters the circle. The basic steps are a quick slide forward and backward while shifting weight from one leg to the other. The upper body of the dancer is relaxed while the hips and legs move.

In 1888, slavery ended in Brazil. Many freed people moved to cities, including Rio de Janeiro. There, they created clubs called samba schools. The schools were a place to **socialize** and dance. Each year, the samba schools competed in a big celebration called Carnival. As the samba dancers performed, huge crowds gathered to watch them. Samba soon became the national dance of Brazil. By the 1940s, it was hugely popular around the world.

Singer and actor Carmen Miranda helped make samba music and dance popular.

Capoeira is both a dance and a fight.

Capoeira is a Brazilian dance—and **martial art**. It was introduced by enslaved Africans. It's a form of self-defense disguised as a dance!

The samba is similar to another Latin dance called the rumba. Like the samba, it **originated** from African dances. The rumba, however, started in Cuba, an island in the Caribbean. Rumba rhythms came from big band music. In the 1930s, the rumba was a recognized ballroom dance.

The mambo, another Cuban dance, replaced the rumba as the most popular Latin dance in the 1940s. After that came the energetic cha-cha. Around 1960, salsa was born in New York City. This flashy Latin dance was created by Puerto Ricans from a blend of Cuban dances. It also included aspects of American swing and tap.

This street art in Seattle shows rumba dance steps.

Rumba dancers in Cuba

Over time, Latin dances continued to **evolve** and gain in popularity. And new styles have come about. TV shows such as *Dancing with the Stars* and *So You Think You Can Dance* helped bring Latin dances like salsa to millions of viewers. Movies and theater have also brought these dances to a wider audience. Today, Latin dance is hotter than ever.

The merengue also hails from the Caribbean. This dance can be traced back to African people in the Dominican Republic and Haiti.

Famous Latin Dancers

"Great dancers are great because of their passion."
—Martha Graham

There are countless Latin dancers who have left their mark on the art form. Rita Moreno is one of the standouts. She was born in Puerto Rico in 1931. Rita and her mother moved to New York City a few years later. When she was a little girl, Rita took dance classes. By 1945, she was performing on Broadway. In 1961, Rita landed a role in *West Side Story*. Her incredible dancing and performance won her an Academy Award—a first for a Latine woman.

Rita Moreno has been performing and winning awards for 8 decades!

Another Latin dance star is Graciela Daniele. Born in Argentina, Graciela has been dancing and directing for nearly 60 years. She thinks dance is a wonderful way to tell stories and connect with people. Graciela started dancing at age six. "I didn't know about storytelling then," she said. "I just loved dancing." After working as a professional dancer, Graciela became a director and **choreographer**. She created works for Ballet Hispánico and other companies.

Graciela Daniele with lighting designer, Jules Fisher

Tina Ramirez, a Puerto Rican and Mexican dancer, founded Ballet Hispánico in 1970. She wanted to create a company that embraced all forms of Latin dance. And she did just that! Tina invited choreographers from all over the world to work with Ballet Hispánico. The company now has nearly 80 dances in its **repertoire**. Tina has been called a "cultural trailblazer."

A performance by Ballet Hispánico

In 2009, Eduardo Vilaro took over Tina's role at Ballet Hispánico. Eduardo is a gay Cuban American who started out as a dancer in the company. Then he formed his own company, Luna Negra Dance Theater. Eventually, he returned to lead Ballet Hispánico. Eduardo **promotes** dance through education and **inclusion**. "I want to make everyone feel that they can come in and be a little Latino," he said.

Eduardo Vilaro

Ballet Hispánico is not only a dance company. It's also a dance school! It trains students in a variety of dance styles.

Get Moving!

"Latin dance has a feel-good factor . . . it's a great excuse to forget your troubles."
—Susan Montero

Dance is already a big part of many cultures across the globe. However, thanks to Eduardo Vilaro and the countless others who helped pave the way, Latin dance is reaching more people. And they are using dance to build community. Every time people dance, they're forming connections with others. They're also honoring and celebrating the history of the dance.

Ricardo Vega and Karen Forcano feel this way when they dance. They are world salsa champions from Santiago, Chile. They are known for their superfast speed, precise **technique**, and gravity-defying lifts. Ricardo and Karen love sharing Latin dance with the world. "For us, it's very important," said Ricardo.

Champion salsa dancers Ricardo Vega and Karen Forcano are also known as Karen y Ricardo. They've been partners since they were children.

"Dance enables you to find yourself and lose yourself at the same time."
—Unknown

You don't have to be a professional to enjoy Latin dance, though. Anybody can join in! There are lessons for people of every ability and age. Did you know that dance is really good for you? It's a great way to stay in shape. It helps strengthen your muscles as well as improves **flexibility** and coordination. Latin dance is not only good for your body; it benefits your mind. It's a great stress reliever and mood booster. And it's a way to express your creativity. You, too, will get absorbed in the high energy and thrilling rhythm of Latin dance. So get out there, get moving—and let's dance!

Going to a performance or competition is another way to experience Latin dance.

Learn to DANCE

Here's how to perform a samba step.

Step 1: Stand with your feet together.

Step 2: Step back with your right foot and take a tiny step forward with your left foot.

Step 3: Slide your right foot forward so it meets the back of your left foot.

Step 4: Repeat this series of steps. But now step back with your left foot and take a tiny step forward with your right foot.

Step 5: Slide your left foot forward so it meets the back of your right foot.

Step 6: Once you know the steps, try doing them faster, while relaxing your hips.

Congrats, you just performed some basic samba footwork!

Glossary

choreographer (kor-ee-AHG-ruh-fer) a person who creates the steps and moves for a dance performance

civilizations (siv-uh-luh-ZAY-shuhns) complex human societies

colonized (KOL-uh-nahyzd) when groups of settlers took control of a place, often by force

cultures (KUHL-churz) the customs, ideas, art, and traditions that make up people's ways of life

evolve (ih-VOLV) to develop over time

flexibility (flek-suh-BIL-uh-tee) the ability to bend

inclusion (in-KLOO-zhuhn) the act of being included

Indigenous (in-DIJ-uh-nuhss) relating to people who lived in a place before the arrival of settlers

intensity (in-TEN-suh-tee) a high level of action and effort

martial art (MAR-shuhl ART) a style of fighting or self-defense

originated (uh-RIJ-uh-nay-tuhd) started

passion (PASH-uhn) very strong feelings about something

professional (pruh-FESH-uh-nuhl) a person who gets paid to do something as a job rather than just for fun

promotes (pruh-MOHTS) to help or encourage

repertoire (REP-er-twahr) the dances that a company performs

romantic (roh-MAN-tik) showing feelings of love

signature (SIG-nuh-chur) something an artist is best known for

socialize (SOH-shuhl-eyez) to spend time with and get along with other people

technique (tek-NEEK) a skillful way of doing something

Read More

BOOKS

Musmon, Margaret. *Latin and Caribbean Dance*. Chicago, IL: Chelsea House, 2010.

Rooyackers, Paul. *101 Dance Games for Children*. Alameda, CA: Hunter House Publishing, 1996.

Thomas, Isabel. *Latin Dance*. Minneapolis, MN: Lerner Publishing Group, 2011.

Learn More Online

WEBSITES

Explore these online sources with an adult:

Britannica: Latin American Dance

International Salsa Museum

Latin Ballet of Virginia: Latin Dance History

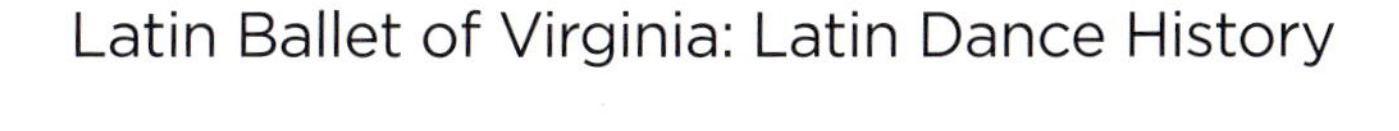

Index

About the Author

Joyce Markovics has written hundreds of children's books. She wholeheartedly thanks Justin Wingenroth for his expert help with this series. Joyce and Justin urge everyone to get up and dance!